Remarks on the Reflection of Conical Shocks; a Memorandum Submitted to the Applied Mathematics Panel, NDRC

point to the second alternative.

Axially symmetric flow and shocks satisfying the basic assumption have been treated by Busemann, by Taylor and Maccoll and by others.* The main objective of those authors was to investigate the flow which results when a conical projectile moves with super-sonic speed. The problem as a whole is explained in a most lucid way by Taylor and Maccoll. As to the analytical formulation of the problem Busemann's procedure is more elegant and flexible. We shall follow Busemann's procedure in the present memorandum.

We first derive shortly the differential equation for the axially symmetric flow assuming that pressure p and density ρ are connected through the adiabatic relation

$$(1) \qquad p = A\rho^{\gamma}, \qquad\qquad \gamma = 1.4.$$

Let x and r be cylindrical coordinates and let u and v be axial and radial components of the velocity. The irrotational character of the flow is expressed through

$$(2) \qquad v_x = u_r ,$$

while the continuity equation is

$$(3) \qquad r(\rho u)_x \quad (r\rho v)_r = 0.$$

By virtue of Bernoulli's equation

$$(4) \qquad \tfrac{1}{2}(u^2 + v^2) + \frac{\gamma}{\gamma - 1} p/\rho \quad \tfrac{1}{2}\hat{c}^2 = \text{const.}$$

and relation (1), one may eliminate the density ρ from equation (3) and

---

* G. J. Taylor and J. W. Maccoll. The air pressure on a cone moving at high speeds. Proc. R. Soc. A. Vol. 139 (1935) p. 276.
A. Busemann. Die Achsensymmetrische Kegelige Überschallströmung. Luftfahrtforschung, Vol 19 (1942) p. 137, and literature given there.

obtains

(5) $(1 - \frac{u^2}{c^2}) u_x + (1 - \frac{v^2}{c^2}) v_r + \frac{v}{r} - 2 \frac{uv}{c} v_y = 0$.

where

(6) $c' = \sqrt{\gamma p / \rho}$

is the velocity of sound. By (4) we have

(7) $c^2 = \frac{\gamma - 1}{2} (\hat{c}^2 - u^2 - v^2)$.

The basic assumption implies that u and v depend only on

(8) $t = x/r$.

Equation (2) becomes

(9) $v_t + t u_t = 0$

while (5) reduces to

(10) $(1 - \frac{u^2}{c^2}) u_t - (1 - \frac{v^2}{c^2}) t v_t + v - 2 \frac{uv}{c^2} v_t = 0.$

The variable t can be eliminated by (9) it is further possible to eliminate differentiation with respect to t and to consider v as a function
of u . To this end we write equation (9) in the form

(11) $t = - v_t/u_t = - v_u$;

Differentiation of (11) with respect to t gives

(12) $1 = - v_{uu} u_t$.

Expressing u_t , v_t , t through v_u and v_{uu}, we obtain from (10)

(13) $(1 - \frac{v^2}{c^2}) + (1 - \frac{v^2}{c^2}) v_u^2 - v v_{uu} - 2 \frac{uv}{c^2} v_u = 0,$

which can also be written in the form

$$(14) \qquad vv_{uu} = 1 + v_u^2 - \frac{(u + vv_u)^2}{c^2} .$$

This is Busemann's equation* (derived in a different manner).

It is clear from (9) and (10) that $u_t \neq 0$ for every solution of (9) and (10) as long as $u \neq 0$. (what happens at $u = 0$ requires special investigation). Every such solution thus leads to a solution of (14) with

$$(15) \qquad v_{uu} \neq 0 .$$

On the other hand, every solution of (14) which satisfies (15) leads to a solution of (9) and (10) when t is defined by (11). In other words, every section of a solution of (14) for which v_u is monotone leads to a possible flow when the ray to which u and v are to be attached is given by (11) and (9).

---

* It may be mentioned that Busemann gives an elegant geometric interpretation of equation (14).

$$(16) \qquad R = N/(1 - \frac{U^2}{c^2}) ,$$

R being the radius of curvature and the meaning of N and U being obvious from the figure.

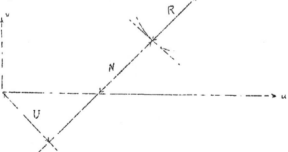

Fig. 4

When v as a function of u is represented by a curve in the (u,v)-plane, the direction of the ray assigned to a point (v,v) is that of the normal to the curve at this point, as seen from (11).

The relations governing the transition through a conical shock are the same as for the plane oblique shock; the curvature of the shock cone does not enter. When the shock cone is a straight cone, as is assumed, the jumps of u,v of p, , and of the entropy are constant along each ray when the basic assumption is satisfied on one side; consequently this assumption remains satisfied on the other side. The flow may continue as a potential flow with constant entropy after crossing the shock. In other words the assumption of proper conical shocks is compatible with the basic assumption.

Suppose a flow, characterized by p_o, ρ_o , u_o , v_o crosses such a conical shock. (It is to be noted that this can occur only if the velocity $v_o = u_o^2 + v_o^2$ is supersonic, i.e. if $v_o > c_o$). The velocity (u_1, v_1) immediately past the shock is located on the loop of a strophoid in the (u,v)-plane. The inclination of the ray which generates the shock cone is perpendicular to the straight connection between $(u_o$, $v_o)$ and (u_1, v_1). The positions of the cones corresponding to the cases (I): $v_1 > v_o$ and (II): $v_1 < v_o$ are indicated in Figure 6.

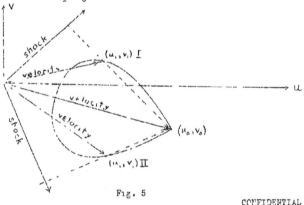

Fig. 5

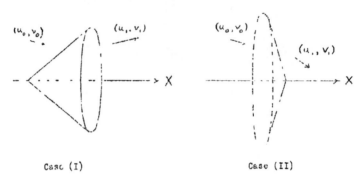

Case (I) Case (II)

Fig. 6

If the state (1) past the flow is prescribed, the points (u_0, v_0) corresponding to possible states in front of the shock are situated on the "tail" of the stroprodi through (u_1, v_1) as indicated in Figure 7 for case (I).

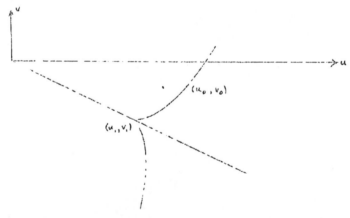

Fig. 7

When the flow on either the front or the back side of the shock is to be continued according to differential equation (14), the slope of the (u,v)-curve is to be so determined that the ray given by (11) coincides with the shock. Since this ray is to be normal to the (u,v)-curve on the

CONFIDENTIAL

one hand and perpendicular to the straight segment connecting (u_0,v_0) with (u_1,v_1) on the other hand, the (u,v)-curve should begin or enter in the direction of this segment. The slope of the (u,v)-curve is thus given by

(17) $\quad v_u = (v_1 - v_0)/(u_1 - u_0)$.

The discussion of conical shocks by Busemann and by Taylor and Maccoll was restricted to case (I) with $u_0 = v_0 > 0$ and $v_0 = 0$. This case occurs when a constant axial flow is deflected by a conical projectile. It may shortly be indicated how Busemann treats this problem. Through the shock transition relations the flow velocity (u_1,v_1) past the shock is given; (observe that the third transition relation guarantees that the Bernoulli constant $\frac{1}{2}\hat{c}^2$ is the same before and after the shock). A solution of equation (14) is to be found whose graph passes through the point (u_1,v_1). The slope v_u of this curve is given by (17). The solution is now to be so continued that $t = x/r$ increases, i.e., in view of (11), v_u decreases up to a point at which the flow and the ray have the same direction, i.e., where $v/u = x/r$, or where the normal passes through the origin; such a point may be called an end point. This end point depends on the choice of the point (u_1,v_1) on the strophoid. The manifold of endpoints that can be reached from $(v_0,0)$ forms a curve which Busemann calls "apple curve" in view of its peculiar shape, see Fig. 8.

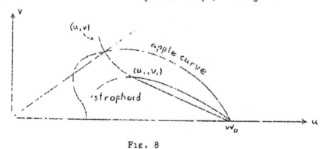

Fig. 8

In this procedure the shock is prescribed and the end direction is found. If the end direction is prescribed one may find the corresponding point on the apple curve by intersecting it with the appropriate ray through the origin. In general there will be two intersections of which the one with the weaker shock is likely to occur in reality parallel to projectile. *

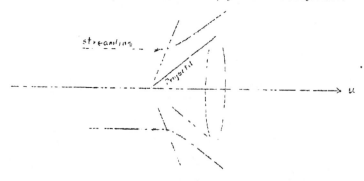

streamline

projectile

u

Fig. 9

We now return to our problem of regular conical reflection. The incident shock is of the first type, the state (0) in front of it being constant with axial flow direction, $u_0 = v_0 > 0$, $v_0 = 0$. The flow past the incident shock (1) will point toward the axis, $v_1 < 0$; it will not remain constant and hence the state (2) in front of the reflected shock will differ from state (1). The reflected shock is of type (II); the flow behind it (3) may either be immediately axial and constant or** it will vary until its direction has become axial and be constant from there on (4), cf. Figure 10.

---

* It may be mentioned that in the procedure of Taylor and Maccoll one is to begin with the end direction; the shock is then found by follow- ing the solution of (10) backwards. This procedure has advantages for single cases are to be investigated.

* This possibility was pointed out to me by v. Neumann.

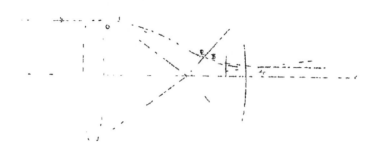

Fig. 10

The flow between the two shocks, i.e. between (1) and (2), (and also that between (3) and (4)) is governed by the differential equation (14) for v as a function of u . The initial direction at (1) (and also at (3)) is determined from relation (17). In order to establish the impossibility of such a configuration, we shall show that the expression $v \, v_{uu}$ is positive past any shock, hence at (1), while it is negative in front of any shock, hence at (2). The sign of $v \, v_{uu}$ must, therefore, change in the transition from (1) to (2). A change of the sign of v_{uu} is forbidden by (15). That v also cannot change sign can be seen as follows: In state (1) we have $v < 0$, hence $v_{uu} < 0$. If v_{uu} does not change sign, $v_{uu} < 0$ when $v = 0$ is reached. Clearly, $v_u \geqq 0$ when $v = 0$ is reached for the first time. On the other hand, $v = 0$, $v_u \geqq 0$ implies $v_{uu} \geqq 0$; this is seen by differentiating (14) and then setting $v = 0$ *. Thus a contradiction is established.

We prove our statement concerning $v \, v_{uu}$ by direct calculation. Let (u_0, v_0) correspond to any state which may be in front of or behind a shock: see Figure 11. Let (u, v) correspond to the state on the other

_____

* The justification of this argument follows from the expansion of $v(u)$ in the neighborhood of $v = 0$ given by Busemann, cf. his (14).

side of the shock. Let (u_*,v_*) be the coordinates of the point (u,v) with respect to a system which is obtained by rotating the u-axis such that it passes through the point (u_o,v_o). Then we have (cf. Busemann)

$$(18) \quad v_*^2 = (w_o - u_*)^2 \ \frac{u_* - w_1}{w_2 - u_*}$$

with the notations

$$(19) \quad w_o = \sqrt{u_o^2 + v_o^2} ,$$

$$w_1 = \nu \hat{c}^2/w_o , \qquad w_2 = w_1 + \lambda w_o ,$$

$$\nu = \frac{\gamma - 1}{\gamma + 1} , \qquad \lambda = \frac{2}{\gamma + 1} .$$

The slope v_u of the solution of (14) which begins or ends with the state (0) is given by (17). Hence we have from (18)

$$(20) \quad v_u^2 = \frac{u_* - w_1}{w_2 - u_*}$$

The quantity

$$U^2 = \frac{(u + vv_u)^2}{1 + v_u^2}$$

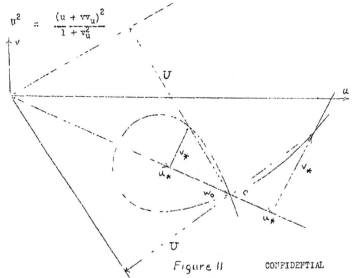

Figure 11   CONFIDENTIAL

is easily determined to be

$$U^2 = \frac{(w_0 - u_*)^2}{(w_0 - u_*)^2 + v_*^2} \, w_0^2$$

or from (18)

(21) $$U^2 = \frac{(w_2 - u_*) \, w_0}{\lambda}$$

In view of (7) and (19) we have

(22) $$\lambda c^2 = w_0 (w_1 - \nu w_0) \, .$$

Hence, from (21) and (22),

$$1 - \frac{U^2}{c^2} = - (w_0 - u_*) \, \frac{w_0^2}{\lambda c^2} \, ,$$

or by virtue of (14)

(23) $$v \, v_{uu} = - (w_0 - u_*) \, \frac{w_0^2}{\lambda c^2} \, (1 + v_u^2) \, .$$

Now, if the state (0) is in front of the shock we have $u_* < w_0$, hence $v \, v_{uu} < 0$; if (0) is past the shock we have $u_* > w_0$, hence $v \, v_{uu} > 0$. Thus our statement is proved.

Lightning Source UK Ltd.
Milton Keynes UK
UKHW020657290721
387974UK00007B/855